I0866836

The INSIDE GUIDE

STEM CAREERS

Jobs in Information Technology

By Beatrice Harris

Cavendish Square

Published in 2024 by Cavendish Square Publishing, LLC
2544 Clinton Street Buffalo, NY 14224

First Edition

Website: cavendishsq.com

Library of Congress Cataloging-in-Publication Data

Names: Harris, Beatrice, author.
Title: Jobs in information technology / Beatrice Harris.
Description: Buffalo, New York : Cavendish Square Publishing, [2024] |
Series: The inside guide. STEM careers | Includes index.
Identifiers: LCCN 2023032253 (print) | LCCN 2023032254 (ebook) | ISBN
9781502670403 (library binding) | ISBN 9781502670397 (paperback) | ISBN
9781502670410 (ebook)
Subjects: LCSH: Information technology–Vocational guidance–Juvenile
literature.
Classification: LCC T58.5 .H379 2024 (print) | LCC T58.5 (ebook) | DDC
004.023–dc23/eng/20230808
LC record available at https://lccn.loc.gov/2023032253
LC ebook record available at https://lccn.loc.gov/2023032254

Editor: Therese Shea
Copyeditor: Shannon Harts
Designer: Deanna Lepovich

The photographs in this book are used by permission and through the courtesy of: Cover, p. 24 Roman Samborskyi/Shutterstock.com; p. 4 fizkes/Shutterstock.com; p. 6 Prostock-studio/Shutterstock.com; p. 7 Rawpixel.com/Shutterstock.com; p. 8 Monkey Business Images/ Shutterstock.com; p. 9 Somphop Krittayaworagul/Shutterstock.com; pp. 10, 14, 18, 21, 22, 23, 25, 27, 29 (right) Gorodenkoff/ Shutterstock.com; p. 12 Joe Techapanupreeda/Shutterstock.com; p. 13 Yurich/Shutterstock.com; p. 15 VTECH Vector and Footage/ Shutterstock.com; p. 16 Thapana_Studio/Shutterstock.com; p. 19 rawf8/Shutterstock.com; p. 20 F8 studio/Shutterstock.com; p. 29 (left) as-artmedia/Shutterstock.com.

CPSIA compliance information: Batch #CWCSQ24: For further information contact Cavendish Square Publishing LLC at 1-877-980-4450.

Printed in the United States of America

CONTENTS

We use computers every day. Laptops, smartphones, and tablets are all kinds of computers.

IT AROUND US

Chapter One

Think about the ways you use computers each day. Much of your time using computers likely has to do with information. That information, or data, might be in the form of a friend's phone number on a smartphone. It might be a photo that you take and share with your family. It might be a show that you **stream** on a tablet. The creation, storing, exchanging, and **retrieval** of information using computers and other digital devices is called information **technology**, or IT.

What Is IT?

IT is a bit hard to explain because the term covers so much. Basically, it's the use of computer systems and other devices to collect and share data. IT workers are those who either construct the systems, networks, and **applications** that handle the data or oversee and maintain these things in some way.

> **Fast Fact**
>
> In computers, all information is translated and stored as digital data, which is a sequence, or certain arranged order, of 1s and 0s.

IT is involved in so many parts of our lives—from saving a paper for school to looking up information about a restaurant you want to visit.

If you play an online game in which your data is stored, you're using information technology.

Fast Fact

IT workers must understand how hardware and software work together. If one piece isn't compatible, or matched correctly, with another, something will go wrong. **Troubleshooting** is a big part of many IT jobs.

If you ask a question of a digital assistant, such as Apple's Siri or Amazon's Alexa, your voice is turned into stored data. The assistant returns an answer using IT too. IT is also involved when we buy things online. Even a trip to the doctor involves the use of information technology for your medical records. IT systems

HARDWARE AND SOFTWARE

This book mentions hardware and software in IT workplaces. So, let's get the difference between hardware and software straight! Hardware is the physical parts of a computer, or the things that you can touch (or that you could touch if you opened up a device). These include the keyboard, the central processing unit (CPU), and the network **interface** card (NIC), which allows a computer to connect to a network such as the internet. Software is the programs, or sets of instructions, that the hardware uses to perform tasks. For example, utility software, such as software that identify viruses, helps IT workers manage applications and systems.

A computer virus is a code or program that inserts itself into another program, copies itself, and harms data as it does this. It can spread to other computers through shared files or software.

are almost inescapable. It takes many talented people to create and maintain these systems.

Interested in IT?

The U.S. Bureau of Labor **Statistics** (BLS), an agency that studies the labor market and the economy, expects the number of careers in computers and IT will increase by 15 percent from 2021 to 2031. This is an increase of more than 680,000 jobs. That means people who can do these jobs will be in demand in the future. There's more good news for those who pursue these careers: The median salary is much higher than the median salary for all workers. It was about $97,430 in 2021.

Many IT job ads ask for candidates with a college

Paying attention in your STEM classes, or those that focus on science, technology, engineering, and math, can help with your future IT career.

Fast Fact

Some IT workers get the skills they need through technology boot camps, which are education courses that teach certain IT skills.

degree that focuses on computer or information technologies. However, it's possible to get a job other ways, such as through **certifications**. Most employers are looking for employees with experience too. Read on to see which jobs seem to fit your skills and talents and which you can see yourself doing in the future.

Boot camps often offer an accelerated, or faster, way of learning.

Servers are powerful kinds of computers that often can store much more data than the average computer. Many servers are needed for a truly large data collection.

THE BUILDERS

Chapter Two

You're likely familiar with the kind of architect that designs and oversees the construction of buildings. In the world of IT, architects create digital structures that hold information. These structures need to be strong and accessible, just like buildings need to be.

Database Architect

Can you picture a database? You might think of a chart you'd make in class. That's not incorrect, but a database is most often seen as a way of organizing data electronically so it can be easily stored and **accessed** when needed. Businesses need databases to store customer information, health-care organizations need databases to store patients' information, and schools need databases to store students' information. Databases are a critical part of the online world.

> **Fast Fact**
>
> Sometimes data in a database is transferred into what's called a data warehouse. This is a kind of electronic storage for data that is used for analysis, or close study.

A database architect is the person who designs a database. They do this according to the specific needs of their employer. They must be able to work closely with people in an organization to understand what

A database with a lot of different kinds of information can be challenging for a database architect to build, but some love a challenge!

data will be stored and retrieved. After the database is built, the architect may maintain it too.

Database architects often have a college degree in either a computer science, data science, or information technology field. They should also be knowledgeable about database development and management software.

Fast Fact

One of the largest databases in the world is the World Data Center for Climate (WDCC). It has 220 **terabytes** of climate research data.

Computer Network Architect

A computer network architect is one of the highest paid computer professionals. This expert designs and builds networks for data

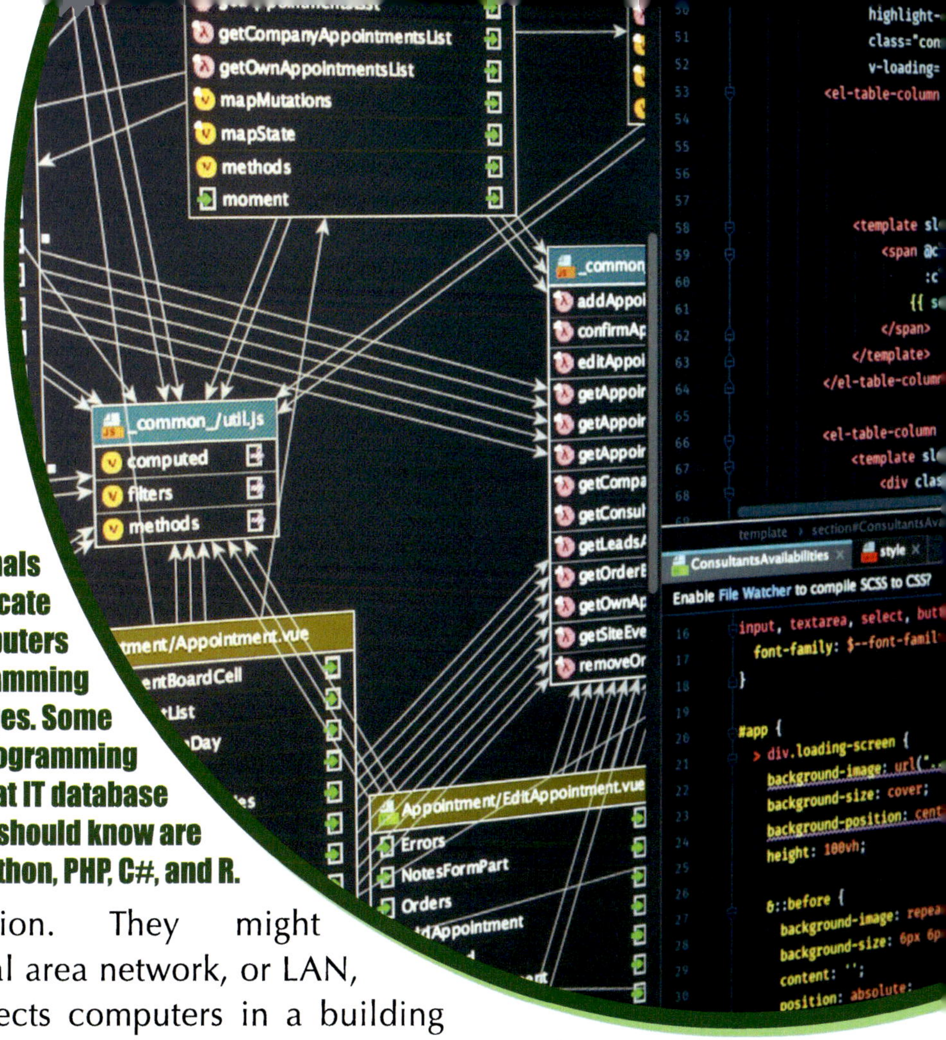

IT professionals communicate with computers using programming languages. Some common programming languages that IT database architects should know are SQL, Python, PHP, C#, and R.

communication. They might create a local area network, or LAN, which connects computers in a building or in an area. LANs can involve a few users, such as in a home, or thousands of users, such as at a college. Wide area networks, or WANs, cover a larger area than LANs and are, in fact, made up of LANs that connect to each other.

Computer network architects meet with their employers to figure out what kind of network is needed. They need to be knowledgeable about the software and hardware needed to build the network. They also need to understand how the network will interact with other software used by their employer. Another area of concern is making a network scalable, a term which means able to expand to meet growth.

ADMINISTRATORS

Once a database, network, or system is built, an administrator is often in control of day-to-day maintenance. They need to address any issues users have and be able to work with the architect or others to add features to address these issues. Additionally, they may need to back up data, which means store a copy of the data somewhere in case the original data is lost. For databases, these people are called database administrators. For networks and systems, they may be called network and computer systems administrators. These careers often require a college degree, but sometimes a related certification and knowledge of the software that architects use is enough.

Administrators act as managers. While they didn't create the product, they're responsible for its smooth operation.

WAN: Wide Area Network

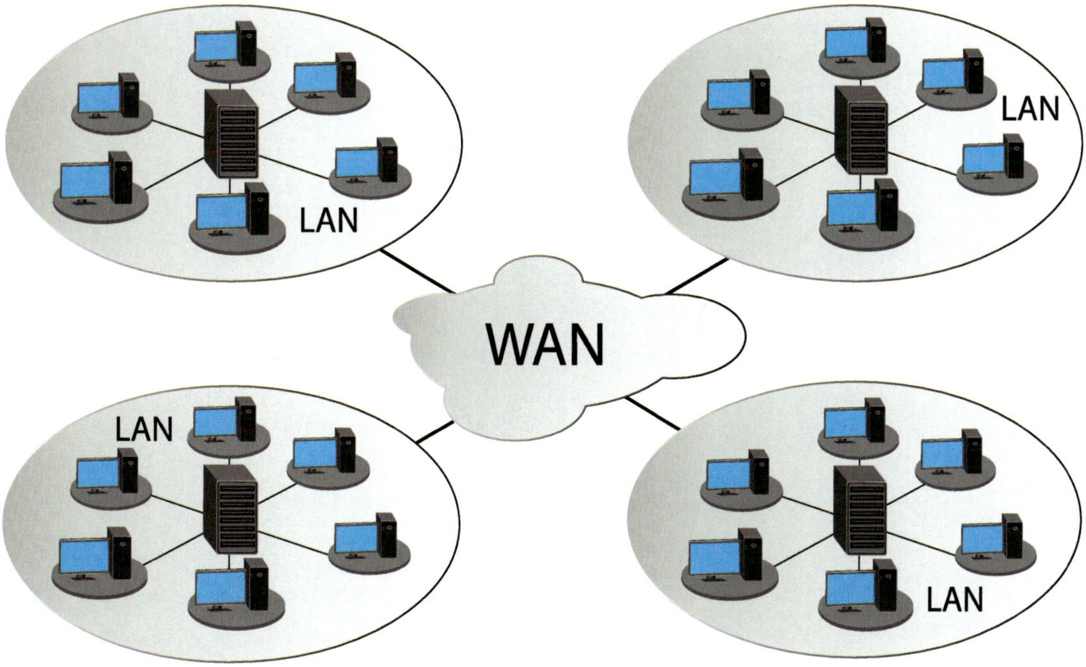

The internet is the largest WAN. It's many, many LANs that can communicate.

After the network is created, they'll help maintain it and troubleshoot any problems.

Computer network architects work for businesses, governments, hospitals, and other organizations. For larger networks, they may work with engineers and administrators. This job may call for working outside of normal business hours or assisting during a network emergency.

Fast Fact

A computer systems architect is another IT career involving hardware, software, and computer networking. A system is a group of computers and programs working together for a purpose.

Cybersecurity is the practice of protecting computer systems, networks, and programs from unauthorized access.

Chapter Three

CYBERSECURITY

Keeping our personal information secure and away from people who may use it for the wrong reasons is important. People called hackers try to get past security programs to steal information from others. They may misuse this information themselves, such as by using someone's bank account numbers to steal money, or they may sell the information to others.

Luckily, there are IT professionals who work in cybersecurity. They strengthen digital security to protect people's data.

Information Security Analyst

Information security analysts analyze, or closely examine, networks and computer systems to find errors or weaknesses in their security programs. They also find ways to strengthen them. In addition, they may monitor them to catch suspicious activity. If a **security breach** does happen, they trace its source and figure out what kind and how much data may have been exposed.

> **Fast Fact**
>
> The BLS reported that the median salary for information security analysts was $102,600 in 2021.

Information security analysts may also act as teachers, educating other

Information security analysts' tools include special software that can help them find threats.

employees and computer users about ways they can safeguard data. This may include information about protecting passwords.

According to the BLS, information security analysts are in demand. It's thought that 33 percent more will be needed by 2030. Information security analysts usually need at least a computer-related college degree to enter the field. Employers often like to see **internship** experience, too, so they know the analyst has some on-the-job training. Information security analysts can also seek

> **Fast Fact**
>
> A cyberattack can cost a company millions of dollars because of stolen money and government fines, in addition to losing the trust of the people whose information was compromised.

WHAT IS A CYBERATTACK?

You've likely heard the term cyberattack before. A cyberattack is any attempt to steal, reveal, change, or destroy data or other digital property in a system, network, or computer. These are the kinds of actions that security analysts are fighting against. People called threat actors attempt cyberattacks for a number of reasons, including for money or as an act of war. They may or may not be skilled enough to be called hackers. One type of attack uses ransomware. This is a kind of malware, or malicious (harmful) software, that threatens bad actions against data unless a ransom, or payment, is given to the threat actors.

Fast Fact

In 2016, the Yahoo company reported a cyberattack compromised, or exposed to danger, the accounts of 1 billion users. It was later revealed the number was actually 3 billion.

The victim of a ransomware attack may see a screen like this. Some threat actors use ransomware to encrypt data, or turn it into a code, that can't be read without a key.

certain certifications, such as Certified Information Systems Security Professional (CISSP), Certified in Risk and Information Systems Control (CRISC), and Security+ certifications.

Ethical Hackers

Not all hackers are bad! Ethical hackers (sometimes called white hat hackers) are people who are employed to find the vulnerabilities, or weaknesses, in a computer system or network. In a way, they pretend to be malicious hackers who are trying to gain access for unethical, or wrong, reasons. When they find a weak point, they work with others

Some hackers use malware to attack computer systems. Malware called spyware operates in secret, gathering data and sending it to hackers.

People should never try to hack into a computer system for fun. The penalty for some hacking crimes is years in prison.

to strengthen it through a software patch, or correction, or another means. Organizations and companies with especially sensitive data, such as banks and government agencies, hire ethical hackers to test their systems. Some pay their ethical hackers more than $150,000 a year.

Those interested in this kind of career should follow the path of other IT professionals by getting a computer science or similar degree in college if they can. Look for certifications, such as the Certified Ethical Hacker (CEH) program, for some experience and focused study. There are even boot camps for people interested in ethical hacking. These IT experts have become a valued member of computer security teams.

Some IT support specialists are called help desk support specialists. They troubleshoot and fix technical problems by providing support via chat, phone, or email.

SUPPORT SPECIALISTS

Chapter Four

Do you like to talk with people? Do you like to help them out when they're having trouble? If the answer to these questions is "yes" *and* you have an interest in technology, there's an IT career path for you! Many IT systems, networks, databases, and other areas need support specialists, or people whose job is to solve users' problems.

IT Support Specialist

An IT support specialist is like a detective. When an IT user runs into a problem, they'll likely speak with a support specialist first. They'll explain their issue. The IT support specialist uses these details, plus their knowledge of the software and hardware involved, to figure out where the problem is and make the corrections needed to help the user.

IT support specialists may help people remotely or in person.

An IT support specialist should know about the hardware in a computer and the software that works with it.

Fast Fact

Many workplaces that rely on technology need 24-hour IT support. Support specialists may work overnight so someone is always available to help.

Besides troubleshooting, IT support specialists also update programs to fix any bugs and install new programs as they become available. For example, antivirus programs need to be regularly

DATA IN THE CLOUD

More and more data isn't being stored where a business, organization, or person is located. It's stored in the cloud. The cloud is the name for the collection of servers around the world that can be accessed using the internet. One advantage to using the cloud is that the business or organization only has to buy storage for their data. They don't have to pay for hardware and software to keep the data themselves. Cloud storage offers many services that can be accessed remotely. Another advantage is that data is generally less likely to be damaged or destroyed in the cloud.

Fast Fact

IT support specialists called cloud support specialists manage an organization's cloud-based applications and services.

The cloud requires a huge number of servers all over the world.

upgraded—or exchanged for newer, better products—to deal with the latest threats. New hardware may also have to be installed as older hardware becomes incompatible with the latest software. IT support specialists may also have to teach people how to use certain software, databases, networks, or systems. In short, they have to know a bit about everything they might be asked about.

IT support specialists work for technology companies, helping people who are having issues with their products. They may also work for other kinds of businesses, governments, schools, or any other organization that relies on IT.

IT support specialists can specialize, or focus, in certain areas too. For example, desktop support specialists specialize in desktop computers, laptop computers, and the devices they connect to, such as printers. There are also database support specialists and security support specialists.

Future IT Experts

IT experts of the future have their choice of industries and organizations to get involved in. People with the careers in this book started with a STEM interest. You can start your career path today by studying especially hard in math and science classes. Join a technology club at school to pick up new skills, such as coding. Also, if your digital devices need repairs, be sure to ask the repair person questions to better understand any software and hardware fixes they make.

Fast Fact

Some expect that, by 2025, the cloud will hold about 50 percent of all the world's stored data.

Young people who learn to code open themselves up to future opportunities for IT success. Coding is writing instructions for a computer.

Finally, keep up-to-date about information technology issues you hear in the news, such as security breaches. The IT experts that can come up with solutions to serious problems like these are in great demand now and will likely continue to be in the future!

IMPORTANT IT SKILLS AND TASKS

examine networks, systems, or devices and make decisions about how well they meet users' needs

aid other staff in troubleshooting and solving problems

back up data when required

install and observe security and other programs

set up computers and software for employees

install upgrades in hardware and software when needed

teach employees about the best uses of technologies

run diagnostics, which are programs that check and test computer hardware or software for problems

THINK ABOUT IT!

1. Why do you think IT careers are expected to grow so quickly in the future?
2. Besides the places mentioned in this book, where else could an IT professional work?
3. Can you think of disadvantages to more information being moved to the cloud?
4. Which IT career appeals to you the most, and why?

GLOSSARY

access: To retrieve, or get back, or to be able to use.

application: A computer or smartphone program that is made to perform certain functions, or tasks. Sometimes called an app.

certification: An official approval given by an organization that says the receiver has the knowledge and skills needed to do a certain job.

interface: A term that describes the connection and interaction between a user and a computer's hardware and software.

internship: Supervised practical experience to prepare one for a future job.

retrieval: For computers, the act of reading data in storage and returning it to a program as needed.

security breach: An event in which someone gets into a guarded place to access secret or private information.

statistics: A subject related to collecting and studying numbers to show information about a topic.

stream: To transfer data in a continuous way, meant to be watched or listened to immediately.

technology: A method that uses science to solve problems and the tools used to solve those problems.

terabyte: A collection of about 1 trillion bytes. (A byte is eight bits, and a bit is the smallest unit of data, either a 0 or 1.)

troubleshoot: To find and fix problems in computers, software, or other applications and devices.

FIND OUT MORE

Books

Duling, Kaitlyn. *Underground Storage Vaults: Protecting Priceless Information*. North Mankato, MN: Capstone Press, 2020.

Morkes, Andrew, and Shaina Carmel Indovino. *Women in Information Technology*. Hollywood, FL: Mason Crest, 2022.

Stratton, Connor. *Great Careers in Technology*. Lake Elmo, MN: Focus Readers, 2022.

Websites

Computer and Information Technology Occupations
www.bls.gov/ooh/computer-and-information-technology/home.htm
Read a list of technology careers on the BLS website, including the median salary for each.

How Does the Internet Work?
www.wonderopolis.org/wonder/how-does-the-internet-work
The internet is how many people get information. Check out in detail how it works.

Technology and Invention
kids.britannica.com/kids/article/Technology-and-Invention/353296
Find out how technology has changed through the years.

Publisher's note to educators and parents: Our editors have carefully reviewed these websites to ensure that they are suitable for students. Many websites change frequently, however, and we cannot guarantee that a site's future contents will continue to meet our high standards of quality and educational value. Be advised that students should be closely supervised whenever they access the internet.

INDEX